Weathering the STORM

by Barbara A. Donovan
illustrated by Rosanne Kaloustian

AF870499

HOUGHTON MIFFLIN BOSTON

It was a warm August afternoon on the coast of Maine. Dara proudly held the mainsheet of a small sailboat. With her right hand she gripped the tiller. This was her first time sailing solo. After taking sailing classes at the marina near her home for the past two summers, she finally had gotten permission from the instructor to go out on her own.

"Don't sail beyond the first group of islands just outside the cove," the instructor had cautioned. "Storms can come up quickly at this time of year."

No chance of a storm today, Dara thought. Wispy tendrils of clouds filled the sky, but the sun was bright and warm. A steady breeze helped her sail quickly outside the cove, close to a group of rocky islands — the kind that dot the coast of Maine like chocolate sprinkles on an ice cream cone.

Dara sailed lazily back and forth between the islands for a while, watching the seals and cormorants sunning themselves, and dreaming about further adventures. *Someday I'll explore one of those islands*, she decided.

Suddenly, from off the Atlantic, a three-foot wave lifted her small boat and sent it careening sideways. The sail flapped wildly. Startled out of her daydream, Dara pulled in the mainsheet and brought the boat about, avoiding a disaster. *Where did that monster come from?* Dara wondered.

She searched the water and sky for the source of the building waves. *Just my luck to get caught in a storm my first day out alone*, thought Dara. Clouds that had seemed harmless earlier were now dark and menacing. The sky in the northeast had an eerie glow.

Dara could see a curtain of rain falling on a group of islands a few miles to the north. She estimated that the storm was about half an hour away. She would need every minute of that half hour to sail back to the safety of the cove. *I can do this*, Dara told herself. But she was scared. The rising waves made her boat bob up and down like a painted horse on a merry-go-round.

Then lightning flashed nearby. *The storm is too close!* Dara thought. *I'll never make it home before it hits!* She tightened the buckles on her life jacket and pulled the mainsheet in farther. She wished she had never set out on this adventure. Why hadn't she invited a friend from sailing class to come with her? Or better still, her mom, who always knew what to do in an emergency? Her mom had even *asked* if she could come, and Dara had told her no! She'd wanted to prove she could do it alone.

But this was not a time for wishful thinking. The approaching storm demanded Dara's full concentration. As the waves mounted, she couldn't predict where they would come from. A sudden five-foot wave spun the boat around, and the tiller slipped from her hand. The boom swung dangerously close to her head. Ducking, she grabbed the tiller, which rocked crazily back and forth. Gradually she brought the boat under control and began sailing in tight circles. The circles, she knew from her sailing classes, would keep her in about the same place while she thought of a plan. The marina was too far, she decided. Her only chance was to try to land at one of the nearby islands.

Problem was, they weren't so near anymore. The rising wind and choppy waves had pushed her about a quarter of a mile away from the rocky island she'd sailed past earlier, where she'd watched the seals and ignored the darkening skies. Still, Dara was confident she could reach the island again before the storm hit full force. She set a zigzag course and willed herself to think positive thoughts.

Dara was too scared to look back at the advancing storm. An unsettling darkness spread across the sky, blocking out the sun. With the wind whipping around her, she concentrated on holding the boat on a steady course. She tried to remember everything she'd learned in her two years of sailing, wishing she had paid closer attention to the emergency instructions.

After endless tacking back and forth, Dara brought the boat near the rocks, now empty of seals.

Lucky seals, thought Dara. *They can just swim away from the storm.*

She searched the rocky shore, looking for a place to land. A burst of lightning illuminated a patch of sand, no bigger than a large beach towel, between the rocks. It was her only hope.

Dara headed toward the small beach but had trouble controlling the boat. Then, just twenty yards from shore, a large wave drove the boat directly onto some jagged rocks, gashing a hole in the boat's hull. Water gushed into the bottom of the boat and swirled around Dara's feet. Desperate, she decided she would have to swim for shore.

Just as Dara was about to jump into the angry ocean, another wave lifted her boat off the rocks, throwing her off balance. She fell backward, then gripped the side of the boat tightly as the storm drove the boat toward shore. Suddenly, with a dull thud, the boat grounded in the hard sand. Dara shook with relief. She was bruised but safe.

Before she could check the damage to the boat, a monstrous clap of thunder exploded overhead. The rain was about to hit. Dara dragged the boat above the seaweed that outlined the farthest reach of the tide. She stripped off the sail just as the first drops of rain hit the sand.

Dara leaned the boat against two sturdy boulders and tied it to a tree. She plugged the hole in the hull with a boat cushion. Just as she was about to duck under the boat, she remembered her mom telling her to stay off the damp ground. The sail would make a good ground sheet. She struggled to spread the stiff, heavy cloth under the boat. It was beginning to get dark by the time she finally crawled into her temporary shelter, exhausted and soaking wet.

She reached for the plastic emergency box in its cubbyhole under a seat. Inside she found several bandages and a roll of gauze, some first-aid ointment, a whistle, a small flashlight, a book of matches, and an energy bar. Dara laughed with relief. The snack bar was her favorite kind.

At least she would have something to eat. But what about water? Her mom had told her that people could survive without food longer than they could survive without fresh water. Dara emptied the contents of the emergency box and rinsed it with rainwater, wiping the inside with the damp edge of her shirt. Then she placed the box outside the boat to catch the rain.

The rain drummed loudly on the boat's hull. Dara shivered in the cold Maine night. She found a dry windbreaker tucked under the boat's bow and slipped it on, but it offered little protection from the night air. Dara tried to wrap part of the sail around her to keep warm, but it was hard to cover her shoulders and legs with the rigid cloth. At the very least, the sail would keep out the wind and rain.

Gradually, the storm moved out to sea. The sound of rain and wind was replaced by the peaceful sound of waves falling over the rocks. Dara peered out into the dark night. The sky above her was an inky black bowl filled with silvery stars. It would be a clear day tomorrow, she predicted. But with a hole in the hull of her boat, she couldn't sail home. In fact, she wasn't even sure which way home was.

Dara eyed the snack bar hungrily. It was all she had to eat, and she couldn't scout for food in the dark — what if she fell and broke her ankle, or put her hand into a patch of poison ivy? Oh, that would be just perfect! She decided to eat half the snack bar now and save half for the morning. When daylight came, she might find wild blueberries and raspberries. But then what?

As Dara munched tiny slivers of the energy bar, she thought longingly of home. By now her mom would have called the harbor police or the coast guard. The storm would have made it hard to conduct a search, and the police probably wouldn't even try at night. She wondered how long she'd be able to survive on the island and what the chances were that someone would find her the next day. Maybe she could make some sort of attention-getting signal to alert them.

Dara pulled the stiff sail close around her and tried to sleep. For a while it seemed like she would never get comfortable and would never stop her mind from reeling with worries. Just when she had finally nodded off, a soft scuffling noise woke her. Could there be bears on this rocky little island? Dara's heart pounded furiously at the thought. She slowly reached for her flashlight. Flicking it on, she saw crabs hustling along the wet sand, close to her head. "Bears!" she said with a laugh. Dara listened to the crabs until she dozed off again.

Within minutes, it seemed, a loud barking startled her awake. Disoriented, Dara bumped her head against the boat and struggled to get out of the tangled sailcloth. With a sigh, she realized where she was and sank back against the hull. The barking continued. It was the seals. They had come back to their rock. Dara remembered how happy she'd been to see them just yesterday afternoon. It seemed a long time ago now.

The sky was beginning to lighten. Dara gave up on sleep and got up, stretching her sore, weary body in the early-morning light. She devoured the other half of her energy bar and watched the rising sun turn the sky pink and gold. It promised to be a beautiful day.

Or it *might* be a beautiful day if only she could get home.

"Let's see. The sun rises in the east," Dara said aloud. The storm had blown her into unfamiliar water, but Dara knew that if she sailed west, she would find the mouth of the bay. For a moment she felt hopeful, but then she remembered the hole in her boat. "If only this were a fairy tale or a dream," she said to the seals on the rocks. "Then one of you guys would give me a ride home."

Still hungry, Dara searched the bushes for blueberries and found a handful. While she popped them into her mouth, she scanned the horizon. If she stayed out in the open, near the seal rocks, a passing motorboat was bound to spot her — wasn't it? She put on her bright-orange life jacket to improve the odds. Then she decided to spread her boat's sail out over the rocks. That would get attention — wouldn't it? As she heaved the stiff, heavy sail over some boulders, she felt dizzy

with hunger and exhaustion. She plunked herself down on a rock and waited. And waited.

Over the next hour, she saw several boats, all of them far away. Orange lifejacket or no orange lifejacket, sail or no sail, they'd never notice her unless they had binoculars trained directly on "her" island. Maybe she should make a fire! She jumped up excitedly and ran to the boat. Crawling under the hull, she found the book of matches. Soggy!

Then a patrol boat sped by, close to shore. Dara jumped up and blew the emergency whistle with all her might. She stood on a high rock and waved her arms over her head.

The boat passed on. Dejected, Dara glared at the seals.

"You guys are barking too loud," she complained. "The patrol boat couldn't hear my whistle." She slumped down on a rock and put her head in her hands. Now what?

Then the sound of a motor made Dara look up. The patrol boat had veered around in a huge arc and come close to shore. Dara's mother was sitting in the bow of the boat, next to a coast guard officer. "Boy, am I glad to see you!" shouted Dara as the boat landed. She realized how scared she'd been when she felt relief wash over her body.

Her mother jumped off the boat and ran to her. "I just knew you'd be okay," she said, giving Dara a gale-force hug. The hug told Dara how worried her mom had been. "You've got a true sailor's instincts, Dara!" her mom said. "Congratulations on weathering your first storm!"